Advancements in Robotics

How AI is Creating a New Era for Robots

Table of Contents

Chapter 1. Introduction

In this special report, we delve into an intriguing yet grounded exploration of the fascinating world of robotics. Intersecting technological advancements and our daily lives, "Advancements in Robotics: How AI is Creating a New Era for Robots," invites readers to join an engrossing journey demystifying the complexities of this extraordinarily relevant field. We uncomplicate the technical jargon, presenting the information in a manner that is both digestible and engaging for the lay reader. This report doesn't just stay confined within the realms of high-tech science labs, but widens its scoop to encompass how these robotic advancements will impact everyday life, businesses, and global dynamics. With an exclusive insight into industry narratives and expert commentary, this special report is an essential guide for anyone interested in understanding the extraordinary marriage of robotics and Artificial Intelligence that's shaping our future. Dive in, and unlock a new appreciation for the world of tomorrow, being built today.

Chapter 2. Unveiling AI: An Introduction to Artificial Intelligence in Robotics

Artificial Intelligence, often referred to as AI, is a term that emerged from the fiery forge of science fiction and made a grand leap into the realms of technology, innovation, and everyday discourse. Like an unsolvable enigma shrouded in mystery, its definition and implications often stir up more questions than they answer. This initial section of the chapter aims to introduce AI and clarify these very intricacies, undergirding the conversation about its role in robotics.

2.1. Defining AI

AI is, at its core, the simulation of human intelligence processes by machines, particularly computer systems. These processes encompass learning, reasoning, problem-solving, perception, and language understanding, among many others. The key factor distinguishing AI from robot behavior not considered "intelligent" is the ability to respond to different scenarios in a non-pre-determined manner or to learn from those scenarios and enhance subsequent responses.

2.2. Types of AI

There are broadly two types of AI – narrow or weak AI, and robust or general AI. Narrow AI is what we see most around us – machines that can perform specific tasks as well, if not better, than humans. This takes forms as diverse as suggesting a movie on Netflix, to identifying abnormalities in an X-ray.

On the other hand, General AI is truly capable of understanding, learning, and implementing knowledge across a multitude of activities, akin to human intelligence. This might seem more exciting and futuristic, but we are a long way from realizing it entirely.

2.3. AI in Robotics

Now that we've peeled back the surface layer of what constitutes AI, let's delve into the meaty subject of how this powerful tool is reshaping the landscape of robotics.

AI integration within robotic systems introduces a new game plan: robots are now not just programmable systems performing a rigid set of tasks. They are getting upgraded to solutions capable of thinking, adapting, and making decisions. That's the power AI brings to the table – the ability for robots to emulate human-like intelligence and cognitive functions.

2.4. AI-based Robotics: Key Advances

AI has led to multiple innovations in the robotics landscape, a few shining stars of such advancements include:

1. Autonomous driving: AI empowers self-driving vehicles to navigate and make real-time decisions on the road.

2. Advanced manufacturing: Numerous industries now leverage "smart" robots to optimize manufacturing processes.

3. Healthcare and surgery: AI-enabled robots are revolutionizing everything from patient care to performing sophisticated surgeries.

These applications merely scrape the surface of the ocean of possibilities that AI-based robotics can open. The following sections will delve deeper into them, exploring the impact and potential of AI-based robotics.

2.5. The Brain of AI: Machine Learning

One of the significant mechanisms enabling AI's cognitive abilities is Machine Learning (ML). Essentially, ML is a system's ability to automatically learn and improve from its experiences without being explicitly programmed for those experiences. It's akin to teaching an infant about the world, where the infant is the robotic system, and its experiences are data.

This discussion wouldn't be complete without the mention of Deep Learning, a sub-field of machine learning. Deep Learning is the closest we've come to emulating the functioning of the human brain. It uses algorithms called "neural networks," designed to recognize patterns, just as humans do.

2.6. Final Thoughts

From disembodied Siri to the metallic sheen of industrial behemoths, AI is an inextricable part of our mechanical counterparts. It brings forth not only a profound transformation in robotic capabilities but also significant potential for societal impact.

As we head into an era where human-like robots could become our companions, healthcare assistants, educators, or even overlords, understanding AI becomes critical. It not only empowers us to participate in the conversation around ethical considerations but prepares us to navigate a world promptly becoming more synthesized with innovative tech.

The subsequent sections will further dissect the multitude of ways AI shapes robotics, from autonomous systems to interplanetary explorers. Stay tuned for in-depth discussions on remarkable and pivotal advancements, as we wade deeper into the AI's riveting narrative in robotics.

Chapter 3. Sensors and Actuators: The Nerves and Muscles of Robots

Robotics is an interdisciplinary field that draws on the combined application of electronics, mechanics and computer science to create systems capable of substituting human work. At the heart of each robot, making sense of the surrounding world and operating dependently, are sensors and actuators, constituting the nerves and muscles of such systems respectively. Establishing the blueprint and enacting actions, they differentiate static machines from dynamic, autonomous robots. Hence, for a comprehensive understanding of the fascinating world of robotics, a deep dive into these two critical elements is crucial.

3.1. Sensors: The Nerves of Robots

The advent of new sensor technologies has revolutionized the capacity for machines to interact with their environment. As nerves in the human body transmit information about environmental factors to the brain, sensors on robots perform a similar function. They collect nuanced data about their surroundings, converting various physical quantities into signals that a robot's 'brain', or computational unit, can interpret and employ.

Light detection and ranging sensors (LIDAR), for instance, use pulsed laser light to ascertain distances and identify obstacles. This is invaluable in several robotic applications, particularly in autonomous vehicles. Furthermore, proximity sensors, thermal sensors, sonic sensors, accelerometers, force sensors, and a myriad of others contribute to a robot's awareness and enable adaptive interactions with the environment.

Advancements include not just in the breadth of detectable physical quantities, but also sensitivity and precision. Such progress propels the evolution of 'intelligent' robots that can conduct tasks previously thought possible only for humans.

3.2. Actuators: The Muscles of Robots

While sensors empower robots to sense their environment, another key component is required to facilitate action - the actuator. Like muscles in the human body, actuators enable a robot's movement, bringing robots to life - giving them the ability to perform complex tasks. From picking delicate fruit, executing precision surgery, exploring hazardous environments, to performing rescue operations, actuators make all these and more possible.

Actuators convert various forms of energy, typically electrical energy, into mechanical motion. Common types of actuators include motors, solenoids, and pneumatic and hydraulic systems. Motors, for instance, are utilized in robots for rotating joints, driving wheels, or propelling drones.

An actuator's functionality can be enhanced with a gear system to adjust speed or exertion of a desired force. For precision robotic operations, notably in medical surgeries, piezoelectric actuators have been employed. These transform electrical energy into precise, minute linear motion, showcasing the wide range of actuator adaptations possible.

3.3. Integration: Channels of Coordination

While sensors and actuators are fundamental to each robot's functioning, their symbiosis is just as crucial. Sophisticated systems depend on Integration, where sensory information is processed and relayed to actuators, enabling purposeful actions. This dynamic integration resembles the coordinated effort of our own neurons and muscles in responding to sensory inputs.

Consider robotic manipulators widely used in industries, like the automobile assembly line. Here, sensors perceive information about the object's presence and location. The robot's internal system processes this information and commands the actuator. The actuator then induces movement in the robotic arm, ensuring precise interaction with the object. This interactive cycle enables the robot to conduct complex tasks independently and repetitively with remarkable precision and efficiency.

The effectiveness of this integration is instrumental in achieving a robot's autonomy, adaptability, and performance.

3.4. Future Trends

As we step into an era marked by the intersection of AI and robotics, there is rapid evolution in sensory and actuator technologies. The pursuance for better, faster, and more capable robots is driving researchers relentlessly. We are constantly seeing more perceptive and responsive robots being designed to serve in fields like healthcare, agriculture, security, manufacturing, and domestic help, among others.

While the ultimate goal is full autonomy, there is an intriguing trend towards building robots that work in conjunction with humans, termed 'cobots'. Their design stress on advanced sensor and actuator components that not just enable interaction with humans but ensures their safety as well, attesting to the critical role these components play in making robots an integral part of our world.

In conclusion, just as nerves and muscles are indispensable in enabling any of our functionalities, sensors and actuators are lifeblood to a robot. They depict the beauty that is the amalgamation of the fundamental principles of electronics, mechanics, and computer science, which converge on the creation of increasingly competent robots, showcasing once more the extraordinary synergy of robotics and AI. Through understanding these components, we gain an appreciation of how they power these magnificent machines to become functional, autonomous entities, thereby shaping the promising future of robotics.

Chapter 4. Understanding Autonomous Machines: Revolution in Decision Making Abilities

Understanding the complex world of autonomous machines requires the comprehension of how decision-making abilities are revolutionized within these marvels of technology. When you have a machine capable of making independent decisions, it no longer remains a reactive entity, merely responding to commands. Instead, it morphs into a proactive one, characteristically shaping its output based on the myriad of inputs it encounters.

4.1. Understanding Autonomy in Machines

Autonomous machines, primarily, are designed and developed to perform operations, tasks, or actions based on their programmed algorithms and learning abilities without any continuous direct human control. Equipped with sophisticated hardware and software capabilities, these machines can sense their environment, process the amassed data, and make informed decisions based on artificial intelligence (AI) and machine learning (ML). This concept is massively influential within the robotics landscape as it moves towards an autonomous era wrapped in unprecedented efficiency and productivity.

The driving force behind this revolution is Artificial Intelligence. An autonomous machine learns by observing and understanding the patterns, correlations, and relationships within the data it's subjected to. For instance, an autonomous car understands its distance from the vehicle ahead, not due to hardcoded instructions, but through an intelligent learning model programmed into itself.

4.2. The Anatomy of Decision Making in Autonomous Machines

The decision-making process in an autonomous machine is accomplished through a series of steps: data gathering, data processing, decision making, and action. These steps are connected and integrated, forming an 'information loop' that facilitates continuous learning and decision-making capabilities.

Data gathering constitutes the first step. Autonomous machines equipped with varied types of sensors gather information from their environment. You might consider self-driving cars that use radar, LIDAR, ultrasonic, and cameras to constantly observe their surroundings.

After gathering this information, the machine then processes the data, making use of premade algorithms—the software aspect of the machine. The accuracy and efficiency with which an autonomous machine can make decisions hinge largely on the potency of these algorithms.

Post data processing, decision-making comes into play. This process is where the rubber meets the road. The machine applies its AI and ML principles to the processed data, using its stored knowledge and learning to generate output. The result could range from simple outputs like turning on switches to complex choices like deciding the best path in a crowded environment.

In the final step, the machine acts based on its decision. For example, an autonomous drone adjusts its propeller speed or direction based on calculated data to avoid collision. The given pathway demonstrates a simple representation of the intricate decision-making sequence within an autonomous machine.

4.3. The Growing Role of Machine Learning in Autonomous Machines

Machine Learning, a subset of AI, is now an integral part of autonomous machines. By implementing ML algorithms, autonomous machinery can learn and improve from experience, enhancing their decision-making capabilities over time. This learning ability is pivotal for complex situations where preprogrammed responses might not be enough.

ML can be divided broadly into three categories: Supervised learning, Unsupervised learning, and Reinforcement learning. In supervised learning, the machine is trained with labelled data, enabling it to make future decisions based on its training. In unsupervised learning, the machine is left unsupervised to find patterns in the data, a mechanism that is beneficial for clustering or organizing data. Reinforcement learning, on the other hand, rewards the machines for correct decisions, and penalizes for the wrong ones, coercing the machine to learn from its mistakes, similar to training a pet.

4.4. Real-world Applications of Autonomous Machines

Today, the shift towards autonomous machines is apparent in a myriad of industries. Self-driving vehicles are revolutionizing the automotive sector, autonomous drones are reshaping logistics and surveillance, and intelligent machinery is transforming manufacturing.

In healthcare, autonomous robotic surgeons exhibit remarkable precision, reducing human error and enhancing the surgical outcome. These surgical robots use sensors to understand the patient's anatomy and make calculated movements during surgery. The decision-making process involved fixates not on the surgery alone but also on patient safety and surgical success rates.

Customer service has also seen a surge in 'chatbots' - autonomous, AI-powered programs capable of interacting with humans. These chatbots use AI and natural language processing to understand customer queries and provide responses accordingly, mimicking human conversation styles.

Robotic Process Automation (RPA) has invaded corporate sectors, streamlining workflow processes by taking over mundane, repetitive tasks. These RPA bots study existing work patterns, understand the rules of operation, and decide how to perform the tasks based on the gathered insights.

4.5. The Future of Autonomous Machines

The advancements we presently acknowledge are just the tip of the iceberg. The future of autonomous machines, steered by enhanced AI and ML algorithms, promises a host of revolutionary developments. The accessibility, precision, and productivity these machines offer are slated to progressively overhaul life as we know it, crossing the threshold from being mere conveniences to becoming necessities.

In a world increasingly defined by pace, the strength of autonomous machines lies in their efficiency, and their ability to constantly learn, analyze, and make informed decisions. The promise of such autonomous technologies fashions an exciting tomorrow—one where machines become our partners, not our replacements, in the voyage towards an even more technologically realized civilization.

From reality-defining concepts to potential applications, autonomous machines stand at the precipice of a fresh age of technological progress. Their decision-making abilities are rapidly improving, setting the stage for a new revolution, one where the invisible line between humans and machines seems destined to blur further.

Chapter 5. Navigating the World: Robotics and AI in Transportation Systems

The intersectionality of transportation and emerging robotic technologies promises profound changes in the way we commute, deliver goods, and interact with our environment. From self-driven cars, traffic management systems, to drone deliveries, the applications of robotics and AI in transportation are vast, efficient, and environmentally friendly.

5.1. The Age of Autonomous Vehicles

Autonomous vehicles have been the spotlight of discussions around AI and transportation for the last decade. Self-driven cars, enabled by AI, have become a reality thanks to companies like Tesla, Waymo, and Uber, who have pushed the boundaries of what's possible in this field.

It all began with establishing the foundational technology of autonomous navigation. Using complex algorithms, sensor input, and vast databases, AI can now replicate (and in some cases surpass) human-like reactionary driving. This marriage of machine learning and robotics in the form of autonomous vehicles has the potential to revolutionize the way we travel every day.

The vehicle's AI system, often referred to as the "brain" of the car, continuously learns from each trip, making future trips safer and more efficient. In terms of the technology, a combination of radar sensors, cameras, lidar scanners, and ultrasonic detectors form the "senses" of these vehicles, feeding real-time data back to the AI.

5.2. Smarter Traffic Management

AI and robotics also have an integral role in transforming traffic management systems. Machine learning algorithms can analyze patterns in traffic flow and adjust signal timings accordingly, reducing congestion and improving commute times.

For example, Google's AI company DeepMind collaborated with Alphabet's mapping service to optimize real-time allocation of green light timings in response to traffic conditions. Powered by reinforcement learning, the AI model trained in a simulation environment, then implemented to real-world intersections, resulted in a 15% reduction in delays.

More interestingly, the evolution of connected vehicles – vehicles that communicate with each other and with traffic management systems –paves way for a revolution in traffic management. AI manages these interactions, optimizing traffic flow and predicting potential risks, thereby enhancing overall road safety.

5.3. Toward AI-Driven Logistics and Delivery

AI-driven robotic systems have found a substantial foothold in logistics and delivery over recent years. From order sorting in warehouses to final doorstep delivery, these systems aim to increase accuracy, speed, and cost-efficiency.

Drone technologies couple rotary or fixed-wing form factors with advanced AI for package delivery. These machines navigate using GPS, avoid obstacles with onboard sensors, and drop off packages autonomously. Amazon's delivery drone and Wing, a subsidiary of Alphabet, are notable examples of this technology.

In warehouses, smart robotic systems powered by AI optimize

picking, packing, and sorting operations. These machines can navigate complex warehouse settings, selecting items accurately and swiftly.

5.4. Robotics on Rails and at Sea

Robotic advancements have not left the railways and seaways untouched. The next generation of trains powered by AI is anticipated to bring increased efficiency and safety to passenger and freight transportation.

Railways around the globe, such as Japan's SCMaglev and autonomous trains in Australia's mining industry, are meeting safety, speed, and punctuality standards. Robotic systems control momentum, make micro-adjustments for smoother rides, and monitor railway health.

Meanwhile, the maritime industry sees autonomous ships as an avenue for lowering costs and improving safety. With technologies such as advanced radar systems, predictive analytics for health monitoring, and algorithms for optimal route selection, AI and robotics stand to significantly optimize maritime travel.

In conclusion, the amalgamation of robotics and AI in transportation systems offers a new lens to view the future. With continual research and sustained development, AI can assure transportation efficiency and safety, all while furthering sustainability efforts. While there are challenges such as ethical considerations and regulatory boundaries, the intersection of robotics and AI is undoubtedly shaping the future of transportation towards a more efficient, safer, and environmentally sustainable direction.

Chapter 6. AI in Healthcare Robotics: A Future with Robotic Surgeons

The evolution of healthcare technology has taken gigantic strides over the past few decades. Today it stands as an emblem of futuristic transcendence with the advent of Artificial Intelligence (AI) and robotics seeping into myriad facets of healthcare. The sub-domain of surgical robotics particularly has witnessed a seismic shift, from the days of speculative science fiction to a foreseeable reality.

6.1. A Brief History of Surgical Robotics

In the quest to improve surgical outcomes and tackle inherent human limitations, the idea of integrating technology was explored earnestly starting in the late 20th century. Early robot-assisted surgeries were prototypes, learning grounds that often required the surgeon to be part of the procedure itself, by controlling the robot remotely. The groundbreaking 'da Vinci Surgical System' launched in 2000, spearheaded this robotic revolution, contracting the gap between human error and precision with its miniature instruments and unprecedented view inside the human body.

6.2. The Concept of AI in Surgical Robotics

AI is swiftly transforming the scope of what surgical robots can achieve. Unlike traditional surgical robots, which were fundamentally 'master slave' systems completely dependent on human command, AI-powered robots can learn from the data of previous surgeries, simulate scenarios, and even predict the outcomes of a surgical intervention.

Robotic Process Automation (RPA) fused with AI has been a game-changer. RPA, traditionally a stronghold of the IT industry, refers to software bots performing repetitive tasks. When applied to surgical robots backed by AI, this automation not only expedites the overall process but adds a layer of efficiency and precision, reaching areas manually inaccessible thus, raising the standard of care exponentially higher.

6.3. Pioneering Developments and Key Players

Several key players in the market, such as Intuitive Surgical, Medtronic, Stryker Corporation, and CMR Surgical, are consistently innovating to enhance surgical prowess. Medtronic's Mazor X Stealth Edition and Intuitive Surgical's da Vinci SP, for instance, are automated systems designed to provide more precise spinal surgeries. STAR (Smart Tissue Autonomous Robot), another innovative leap, outperformed human surgeons in a soft-tissue procedure, making it a possibly transformative force in the surgical robotic realm.

6.4. Ethical and Regulatory Considerations

The infusion of AI in surgical robotics, while exhibiting massive potential, raises significant ethical and regulatory concerns. The definition of liability in case of surgical errors, privacy of patient data handled by AI systems, and transparency in AI decision-making are some of the conundrums necessitating stringent guidelines.

Data security and privacy are supreme. AI systems thrive on large volumes of data. Ensuring that this data handling doesn't compromise patient confidentiality is crucial, necessitating robust cybersecurity protocols. Regulatory bodies worldwide are consistently amending their policies to keep up with this rapidly evolving technology.

6.5. The Impact on Surgeons and Patients

AI-driven surgery signifies an advancement in surgeons' capabilities, offering a high-definition, magnified, 3D view of the surgical site, and eliminating tremors. For patients, this transition implies lesser trauma, reduced hospital stays, and faster recovery.

Nonetheless, the technology is not bereft of challenges. The training of surgeons for the efficient handling of AI-assisted robots, the high costs associated with robotic surgery, and public apprehension toward an AI-dominated surgical domain are potential roadblocks that need addressal.

6.6. A Glimpse into the Future

Tomorrow's AI in healthcare robotics would strive to take us to an era of 'supervised autonomy,' where the surgical robot does most of the task, and the surgeon supervises and intervenes when needed. Future robots could be equipped with integrated radiology, pathology, and device data alongside the surgical situation's real-time understanding, revolutionizing how surgery is perceived and performed.

AI in healthcare robotics, particularly in surgery, is an exciting realm. Its pace and direction of growth rely considerably on future research, ethical norms, regulations, public acceptance, and our collective vision for a futuristic yet compassionate healthcare ecosystem. The future might be uncertain, but what is undeniable is that the fusion of AI and robotics in surgical procedures will redefine the contours of healthcare, sketching a world of medicine that will be intensely interconnected, supremely intelligent, and incredibly precise. 'A Future with Robotic Surgeons' is not a dystopian speculation. The future, it appears, is already here.

Chapter 7. The Rise of Domestic Robots: Assistant AIs in our Homes

Gone are the days when the term 'robot' channelled visions of cold, industrial machinery bustling in large factories. Today's robots come in smaller, more accessible, and highly beneficial forms, cohabitating with us in domestic surroundings as interactive helpers. Blending Artificial Intelligence (AI) and robotics, these domestic machines are increasingly vital in everyday tasks, bringing a new level of comfort and convenience to our homes.

7.1. The Dawn of Domestic Robots

Artificial Intelligence has breathed life into the once-far-fetched notion of robots for in-home use. Early models were cumbersome and limited, designed chiefly for educational or entertainment purposes. However, rapid advancements in AI and machine learning algorithms have unlocked a new generation of domestic robots. These intelligent machines ranging from Roomba vacuum cleaners to friendly companion robots like the Pepper, are designed to simplify our lives.

Such is the infusion of AI in domestic robots that they are becoming increasingly autonomous, learning from and adapting to their environment. Whether it's maintaining your home, alerting you of a break-in, looking after elderly family members, or just keeping you company, these robots are engineered to be efficient assistants, integrated into the very fabric of our daily routines.

7.2. The Technological Backbone: AI in Robotics

At the heart of this robotic revolution is the sophisticated interplay between machine learning algorithms, Natural Language Processing (NLP), and image recognition capabilities. These technologies empower robots to understand spoken instructions, recognize faces, and navigate their surroundings safely.

Take, for instance, Roomba's line of robotic vacuum cleaners. Robust simultaneous localization and mapping algorithms allow these robots to map a room and plan the most efficient cleaning path, while advanced sensors prevent them from running into obstacles or falling off stairs. These aren't just robots, but learning machines that get better at their jobs over time.

7.3. The Intelligent Assistant Landscape

The development of AI robots extends beyond physical machines. AI-powered virtual assistants like Alexa, Siri, and Google Assistant have become prevalent elements of our homes. These intuitive digital butlers help manage our calendars, answer queries, control smart home devices, provide timely news updates, and even engage in friendly chit-chat.

Advanced NLP makes these interactions possible, allowing devices to understand and respond to verbal commands. Simultaneously, machine learning helps improve the accuracy of responses over time, creating a more personalized experience. As AI technology advances, these virtual assistants are primed to become increasingly proactive, not just responding to requests, but initiating actions based on patterns and predictive behavior.

7.4. The Healthcare Advantage: Robots for the Elderly and Disabled

Healthcare is one area where domestic robots have shown exceptional promise. For the elderly and disabled, bots like the Care-O-bot or Mabu provide invaluable help. Whether it's reminding them to take their medication, alerting healthcare providers in emergencies, or providing companionship, these robots bring a new level of care and security.

Moreover, AI-powered robots can be programmed to keep track of health metrics, identify changes in behavior which could be indicative of a health issue, and facilitate telehealth scenarios. These capabilities promise a paradigm shift in elderly and disability care, ensuring more independence and better quality of life.

7.5. Future Trajectories: Towards a Fully Automated Home

In the foreseeable future, domestic robots could become vital household fixtures. AI advancements like the development of emotion AI, robots that recognize and respond to human emotions, are already on the horizon. Moreover, the convergence of cloud-computing, IoT, and robotics is expected to foster entirely automated smart homes, responsive to personal living patterns.

Despite their promise, domestic robots do pose challenges. Issues like data privacy, cybersecurity, and the need for technical service and maintenance require careful attention. Despite these hurdles, the trajectory is clear: domestic robots are here to stay and will only weave themselves more tightly into the fabric of our daily lives.

In conclusion, the deluge of AI advancements over the years unlocks a whole new arena for robotics, especially in the home setting. The quintessential vision of a robot-infused future is transitioning from merely speculative fiction to reality. As these robots improve, our domestic lives become increasingly convenient and efficient - a testament to AI's potential to redefine the world as we know it. From robotic vacuums trundling across our living rooms, AI-enabled security systems, companion bots cheering up a lonely senior, or a smart voice assistant aiding in our day-to-day chores - the rise of domestic robots is not only imminent, it is already here. How we choose to navigate this new era will have far-reaching implications, lending a sense of urgency to our understanding of this fascinating intersection of AI and robotics. The future, after all, has a knack for arriving unannounced.

Chapter 8. Industrial Reshaping: The Prominent Role of Robotics in Manufacturing

In the shadow of towering assembly lines and amidst the orchestrated clatter of machine parts, the face of industrial manufacturing is being transformed. The leading light of this revolution – robotics, coupled with the sophistication of Artificial Intelligence, is doing more than mere reshaping; it's essentially redefining the manufacturing realm.

8.1. A New Industrial Revolution

The Fourth Industrial Revolution, or Industry 4.0, as it's popularly referred to, is characterized by an unparalleled integration of the digital, physical, and biological worlds. Robotics, at the heart of this revolution, is driving the modernization of traditional manufacturing and industrial practices.

A robot picking up heavy metal parts with precision, an assembly arm applying adhesive perfectly every single time, a cargo drone autonomously delivering components across the factory floor, all underline the compelling role played by robotics in industrial processes. The marriage of AI with robotics has emboldened new capabilities in these metallic workers - machine learning algorithms enable them to learn, adapt and perfect their tasks while predictive analytic capabilities minimize downtime and optimize productivity.

8.2. The Nature of Robots: Beyond the Physical

Understanding the nature of robots takes us beyond simply marveling at the physicality of these devices. At their core, robots are complex systems of sensors, actuators, and computational units embroidering a layer of intelligence on mechanical shells.

This AI-powered intelligence is defined by machine learning algorithms that allow robots to learn from their experience and progressively improve their performance. By coupling this with advanced sensors – from cameras to Lidar sensors, robots gain a contextual, multidimensional understanding of their environments, enabling them to handle objects, navigate spaces, and interact with human beings.

8.3. Implementing Robotics in Manufacturing

The implementation of robotics has seen a broad spectrum of application areas in manufacturing. Repetitive tasks that were prone to human error have now been eradicated, ensuring consistent product quality and increased output.

Drawing on the AI prowess, robots are now engaged in complex tasks requiring adaptation and improvisation. From collaborative robots or 'cobots' working side by side with workers in assembly lines to autonomous mobile robots (AMRs) navigating the warehouse performing pick-and-place tasks, the embrace of robots in manufacturing is expansive.

8.4. Economical Impact of Robotics in Manufacturing

While initial installation costs of robotic systems may seem high, the long-term economic benefits are significant. Robots demonstrate limits that are far beyond any human ability regarding endurance, speed, and precision, leading to increased productivity, improved quality, and decreased waste.

Moreover, with machines handling the dangerous, strenuous or monotonous tasks, worker safety is significantly amplified. In the long run, this translates to cost savings in terms of reduced injury-related costs and increased worker satisfaction.

8.5. Social and Environmental Responsibility

As manufacturers are being held to higher standards of social and environmental responsibility, robotics presents a tangible solution. Robots excel in accuracy, repeatability, and consistency – maximizing resource efficiency and minimizing waste.

In addition, the increased productivity offered by automation exhibits a potential solution to labor shortages and demographic changes, thereby reshaping the socio-economic equations of industrial manufacturing.

8.6. The Road Ahead

The future is already hinting at further integration of AI in robotics, with advancements in technologies like the Internet of Things (IoT), 5G, and Edge Computing paving the way. The increasing adoption of cloud robotics, where collective learning and resource sharing among robots is possible, is promising an intriguing evolutionary path.

Amid apprehensions about job losses, the narrative is slowly transitioning to appreciate the new roles - robot programming, maintenance, and more - that this techno-aided future is bringing forth.

In conclusion, the story of robotics in manufacturing is still being written, but the chapters unfolding are profoundly exciting. It is an era of reshaping and redefining, leading to not just enhanced productivity and economic efficiency but an unprecedented revolution that holds the power to redefine the very contours of industrial manufacturing.

Chapter 9. Artificial Intelligence and Robotics in Defense: Changing the Landscape of Warfare

It is no longer the stuff of science fiction; advanced robotics, powered by Artificial Intelligence (AI), is revolutionizing defense and warfare. From autonomous drones to robotic tanks, AI has significantly altered the norms of combat and revolutionized strategic thinking in a military context.

9.1. The Emergence of AI-Powered Defense

Since World War II, dealing with the potentiality of robotic and automated systems in warfare has been a focal discussion among strategists and military engineers. But the advent of Artificial Intelligence changed the chessboard.

Artificial Intelligence, in this context, refers to computers and software capable of learning and making independent decisions based on data inputs they receive. The true stakes lie in machine learning, a subset of AI. Machine learning allows these algorithms to learn from past experiences, enhancing their capabilities and performance without explicit programming. It marries statistical techniques with sophisticated algorithms to create intelligent robots capable of complex tasks.

Already, we are seeing the novelty of AI play out on the battlefield. Drones are performing surveillance missions without jeopardizing human lives. Autonomous weapons systems can engage targets at a pace far beyond human capability. AI is also streamlining military logistics, from supply chain management to maintenance schedules.

9.2. Autonomous Robots: Eyes and Ears on the Battlefield

Unmanned Aerial Vehicles (UAVs) or 'drones' have become an iconic symbol of the modern battlefield. Initially used for surveillance and reconnaissance, drones can now be armed with lethal weapons to pursue targets. How can they do this? Autonomous technology allows them to analyze vast amounts of data quickly to identify potential hostiles and distinguish friends from foes.

Drones are just the tip of the spear. Underwater autonomous vehicles are being deployed for sea mine detection, while on land, robots perform dangerous jobs, such as IED detection and disposal.

9.3. Robotic Soldiers: Reality or Science Fiction?

Apart from drones and other unmanned vehicles, several countries are already experimenting with more humanoid-like robots. These robotic soldiers could potentially replace human soldiers in high-risk operations. However, it's a double-edged sword. While robotic soldiers could save lives by taking humans out of direct conflict, they also raise ethical questions about AI in warfare.

9.4. AI in Cyber Warfare

Beyond the physical battlefield, AI is also reshaping the cyber warfare landscape. Cyber warfare is a major threat in today's interconnected world, with state-sponsored hackers launching attacks on critical infrastructure or stealing sensitive information.

AI plays a dual role in this sphere. On one hand, it helps bolster defense, detecting vulnerabilities, potential threats, and anomalies in a fraction of the time that human analysis would take. On the other hand, AI can also be part of the offensive arsenal, used to launch sophisticated attacks on enemy networks.

9.5. Ethical Concerns and Regulation

While the potential advantages of AI in defense are enormous, they do not come without significant moral and ethical concerns. The concept of autonomous warfare machines capable of making life and death decisions poses a series of troubling questions about accountability, legality, and humanity in warfare.

The development and use of AI in warfare are under scrutiny globally, leading to numerous ongoing debates on international regulations and treaties to establish "rules" for AI in warfare. As the technology continues to evolve, the semantics of these discussions will evolve as well.

Understanding the implications of AI in defense and warfare is critical and demands broad and mature discussions involving defense experts, technologists, ethical philosophers, and governments in order to avoid catastrophic and irreversible impacts.

Our epoch is on the edge of an AI-driven shift in military technology. The applications of AI in defense and warfare are expanding rapidly, presenting both new capabilities and challenges. The way forward is marked by both tremendous opportunity and considerable risk. As AI continues to infiltrate more military applications, its influence will undoubtedly redefine warfare's landscape. However, it also underscores the urgent need for discussions on the technology's ethical and moral implications. Ultimately, our wisdom in handling this new frontier will determine its impact on the future of humanity.

Chapter 10. Ethical Dilemmas: The Societal Impact and Regulatory Discussions

The dawn of the robotic age brings with it a plethora of questions of ethical, legal, and societal implications. While the science fictions of yesteryears saw these questions coming, it is only now that we are facing them head-on.

10.1. The Moral Quandaries of AI and Robotics

Robots, guided by advanced AI, often appear to make decisions or perform actions with a semblance of autonomy. However, the truth is, these machines are devoid of a moral compass. Every action or decision by a robot is the result of algorithms and programming fed by humans. This raises the question: who is responsible if a robot commits an act that would be considered illegal or unethical if done by a human?

A widely mooted example is the hypothetical situation involving an autonomous vehicle facing a no-win situation on the road. Given the choice to hit a pedestrian or swerve and risk the lives of its passengers, how will the AI decide? More importantly, who will be held accountable for the outcome? This conundrum, known as the Trolley Problem, exemplifies the murky ethical landscape that emerges with the advent of autonomous systems.

10.2. The Impact on Labour Market and Economy

The adoption of robotics is escalating at an unprecedented scale, with EY predicting the sale of industrial robots to grow at a CAGR of 16% between 2020 and 2025. This promises increased productivity and efficiency but looms as a significant concern for many workers. AI and robotics threaten to automate a host of routine tasks, which may render millions jobless. While many argue that technology would create an equal, if not more, number of jobs as it displaces, how the world navigates this transition holds profound societal implications.

According to the World Economic Forum's 2020 Future of Jobs report, by 2025, the number of jobs lost to automation would be 85 million worldwide. However, it also predicts that around 97 million new roles, ones that are more aligned to the new division of labour between humans, machines, and algorithms, would emerge out of the ashes.

While these trends promise a future with more specialized and rewarding roles for workers, the transition is a huge challenge. It calls for robust policies on reskilling and upskilling workers to preserve social equity amidst the influx of robots.

10.3. The Role of Governments and Regulatory Bodies

The regulatory framework for AI and robotics is in its infancy. Global jurisdictions are grappling to establish firm laws for AI, particularly in sectors like healthcare, transportation, and defense. In many ways, the rate of technological progress outstrips the pace at which laws are enacted, creating an urgent need for harmonizing AI and its ethics in legislation.

To create a measured legislation for AI and robotics, a thorough understanding of the technologies and an anticipation of their impact is crucial. This entails collaboration between AI developers, ethicists, sociologists, and policy makers to cover ethical concerns from AI's fairness, transparency, and bias to its accountability and robustness. It's a balancing act, ensuring innovation is not stifled while protecting public interest.

Important efforts are being made towards this end globally. For example, the European Union has established its High-Level Expert Group on AI, which released its AI ethics guidelines in 2019. In the US, the Office of Science and Technology Policy released a draft memorandum for the heads of executive departments and agencies to ensure the regulatory and non-regulatory approaches to AI are based on consistent principles and avoid a patchwork approach.

10.4. The Societal Shifts and Future Outlook

Despite these mounting challenges, the prospects of AI and robotics are too promising to be denied. The societal and economic advantages they bring – from allowing people to delegate mundane tasks, to enabling breakthroughs in healthcare and better decision making in business – cannot be understated.

However, these technologies push the boundaries of what society is comfortable with. As AI becomes more pervasive, it necessitates a shift in our moral and social attitudes. How we embrace this shift depends to a great extent on how well we understand and manage the ethical and societal questions at hand.

The future of AI and robotics isn't something that will happen to us – it's a future we have a hand in creating. By participating in conversations about the ethical implications of AI and robotics, we can help guide these technologies towards creating a future where the benefits are widely shared and the pitfalls are effectively mitigated. The ethical dilemmas of AI and robotics are at the forefront of making this future a reality.

Chapter 11. Future Projections: The Driving Forces and Challenges for AI in Robotics

Advancements in robotics, driven by artificial intelligence (AI), are set to radically transform our world. This revolution will not be without challenges, but the potential benefits are substantial. In this chapter, we will explore the key forces driving this scientific upheaval and the major obstacles that need to be overcome to realize the promise of AI-powered robotics.

11.1. AI and Robotics: The Momentum Gathers

The momentum behind the integration of AI and robotics is generated by several factors, one of which is technological advancements. Ongoing progress in AI, including machine learning and deep learning techniques, is paving the way for increasingly sophisticated robots. These aren't the robot butlers and automated maids of science fiction, but systems capable of solving complex tasks, processing vast quantities of data, and learning from their experiences.

Just as the steam engine powered the Industrial Revolution, AI is the driving force behind the robotic revolution. AI is an expansive field, encompassing diverse technologies that enable machines to mimic human intelligence. Central to AI is machine learning, which enables computers to learn from data without being explicitly programmed. Deep learning, a subset of machine learning focused on artificial neural networks, is responsible for many of the most impressive recent breakthroughs in AI.

Simultaneously, advancements in robotics are providing physical platforms for AI to manifest. Robot manipulator arms, for example, equipped with AI-driven perception and control systems can accomplish a variety of tasks—ranging from complex surgical procedures to intricate assembly lines—once thought to be exclusive to human dexterity and cognition.

11.2. Economic Catalysts Aiding the Integration

The integration of AI and robotics also owes a great deal to economic realities. Businesses and industries globally are recognizing the multifaceted benefits of these technologies - from reducing costs and improving efficiency to enhancing safety and quality.

For instance, manufacturers are using AI-powered robots to automate arduous and dangerous tasks, creating safer work environments and reducing human error. In healthcare, companies are developing robotic caregivers and AI-assisted diagnostic systems to augment human care providers, making healthcare services more accessible and affordable.

11.3. Social Influences Shaping the AI-robotics Intersection

On the societal front, changing consumer expectations are driving the adoption of AI-powered robotics. Today's consumers value convenience, personalization, and efficacy, attributes that technologies like AI and robotics can bring on a larger scale in products and services.

Therefore, in the face of these driving forces, the integration of AI and robotics is not just a possibility—it is a necessity.

11.4. Resistance Forces: Challenges Ahead

While the roadmap for the future of AI in robotics is compelling, it is not without hurdles. The following are significant challenges to be addressed.

11.5. Bridging the Gap between AI and Robotics

Despite the remarkable advances occurring hand-in-hand, AI and robotics remain distinct disciplines with a gap to bridge. Melding the capabilities of AI—like natural language processing, image recognition, and decision-making algorithms—with the physicality and sensorimotor functionality of robots is a complex endeavor. It involves overcoming difficulties pertaining to the robustness of AI algorithms, interoperability, and the fusion of various sensor data.

11.6. Regulation and Ethical Concerns

As robotics infused with AI become more sophisticated and autonomous, questions of regulation and ethics become increasingly pressing. From considerations concerning accountability and fairness to issues of privacy and security—every aspect of AI-driven robotic applications will need to navigate this complex ethical and regulatory landscape.

11.7. Workforce Displacement and Reskilling

The potential displacement of human workers by machines is one of the most pressing sociopolitical challenges related to the rise of AI and robotics. While some jobs may be lost, this technological revolution could also create new types of employment. To prepare for this change, we need to redesign our education systems and implement wide-ranging reskilling programs.

11.8. Technological Constraints

Overcoming technological limitations is another notable challenge. While AI and robotics have come a long way, there are still significant gaps in achieving true artificial general intelligence and creating robust, adaptable robots that can function in diverse real-world environments.

Moreover, developing AI applications that are efficient and affordable on a large scale involves surmounting obstacles related to data requirements, energy consumption, hardware compatibility, and more.

11.9. The Final Word

The journey towards a future where AI and robotics are inseparable will be fraught with challenges. However, the incredible potential of these technologies to improve our lives, enhance our businesses, and create new opportunities cannot be downplayed. By continuing to push the boundaries of innovation, and by facing these challenges head-on, the promise of AI and robotics can become a concrete reality. Herein lies our collective responsibility—to harness this potential, to navigate the challenges, and to shape a future defined by the beneficial co-existence of humanity and intelligent machines.